PRACTICAL

RESTORAT.

STONE
— · AND · —
PLASTER-
WORK

YVONNE REES

WARD LOCK

ACKNOWLEDGEMENTS

Photographs: Jayphotos (pages 7, 15, 18, 19, 23, 54 top, 70); Wood Bros (Furniture) Ltd/Tilbury Sandford Brysson (page 50); Alex Ramsay (pages 58, 62, 63, 67, 75); Barnsley House (page 54 bottom); Mayfair (page 78); Dulux Paints (page 79).

Illustrations: Alan Burton

Grateful thanks to James Kerswell and BCTV for their expertise and assistance in preparing this book.

A **WARD LOCK** BOOK

First published in the UK 1997
by Ward Lock
Wellington House
125 Strand
London WC2R 0BB

A Cassell Imprint

Distributed in the United States
by Sterling Publishing Co., Inc.
387 Park Avenue South, New York, NY 10016–8810

A British Library Cataloguing in Publication Data block for this book
may be obtained from the British Library
ISBN 0–7063–7469 X

Designed and typeset by Anita Ruddell
Printed in Hong Kong by Dah Hua Printing Press Co Ltd

CONTENTS

INTRODUCTION

Nothing looks sadder than a building or garden feature that has been allowed to fall into disrepair and decay: crumbling plaster, chipped stonework, a tumbledown wall – all seem to shout neglect and a lack of love and attention. Unfortunately, it does not take long for any building to reach a serious state of dilapidation. Strong sunlight, winds, frost and damp take their toll annually, and, unless a regular programme of maintenance is undertaken, the slightest breach in the actual fabric of the building soon becomes a major problem. When a building is unoccupied for any length of time, the march of time seems to accelerate and, as damp and cold penetrate through the outer walls or down from a leaky roof, the internal structure begins to crumble too.

You may have a part of your own home that needs serious attention; or perhaps you have missing or damaged architectural period features that you would like to mend or replace to restore the building to its original glory. You may even have bravely taken on a derelict property that needs total renovation, and are wondering where to start. We no longer live in an age where the majority are prepared to pay for the time and expertise of such craftspersons who can execute elaborate plasterwork or painstaking stonework. It is lucky that people do still exist, albeit in small numbers, who have the enthusiasm and talent to demonstrate or pass on the old skills to those who care enough about old buildings to want to spend the time and money on them. Of course, where a feature is of particular historic interest or where the

INTRODUCTION

building has been listed, you are likely to be under legal obligation to restore it accurately. Grants may be available, and it is worth investigating this possibility before you start any work.

Even with a grant, there is not usually any obligation to employ and pay a professional to do the job. If you have the enthusiasm and a little practical skill, there is no reason why you should not tackle it yourself, depending on the level of expertise required. If you are completely renovating your own house, you will probably be keen to do as much of the work yourself as possible anyway. Some of the skills require practice; a few really specialized techniques may need a little professional instruction to give you confidence. Look out for weekend courses, craft workshops, summer schools and night classes, which are usually as enjoyable as they are instructive. Your local education authority or the relevant national craft association should be able to supply you with a list of addresses and venues.

Getting the feature and finish right for a period property will probably require some thorough historical research and close observance to detail on your part. Very old cottages do not usually feature anything elaborate architecturally, but, if you have any plaster or stonework to repair, you will find modern techniques look hopelessly out of place and 'new'. Today's plaster mixes are designed to look smooth and glossy, and the modern master plasterer takes a pride in producing sharp corners and level surfaces that sit uncomfortably with crooked beamed ceilings and uneven floors. Some renovation purists, when dealing with lath-and-plaster walls, will even go so far as to make up the original mixtures with dung and horsehair where sections need replacing. When renovating stonework always try to match the size and colour of the stone as exactly as possible: if you cannot find any lying around in the near vicinity – and you will be surprised at how much you can collect – then make sure it is local or matches any existing parts of the wall. Equally important is the colour of the mortar mix and the style of the pointing between the stones. Attention to details such as these can make or mar a renovation.

INTRODUCTION

When it comes to houses with fancy mouldings, cornices, arabesques, pediments and architraves, or detailed stone carvings, close attention to detail and period accuracy are absolutely essential. If the original is totally missing, then you will have to rely on old pattern books or a reliable reference to get the correct size and style. An even better guide is to visit your neighbours' houses, if they are of a similar age and style, and see if they have any existing features you can copy. You will be amazed how a building recovers its former character once features such as these have been accurately replaced. If you are changing the internal arrangement of rooms in such a house, it is important that any decorative plasterwork is preserved. Resist the temptation to remove it in, say, kitchens and bathrooms; and, if you divide an existing room, avoid cutting across any valuable plasterwork, and continue any mouldings along the new partition walls. You should always try to renovate and repair old existing plaster walls and ceilings or fancy plasterwork and decorative mouldings. Only remove and replace where absolutely necessary.

When the work is done and it looks as authentic as possible, take care not to ruin the effect – and your hard work – with some inappropriate decoration. There may be a temptation to pick out mouldings in different colours, where plain white or a suitable pastel shade, perhaps even a touch of gold, is strictly traditional. Where period style suggests different colours, then do take reliable professional advice. Old-fashioned plaster on walls and ceilings usually looks better in flat paint – a glossy finish shows up every bump and curve. Stonework should never be left exposed on the internal walls of old cottages: whitewash it where necessary. Sometimes a specialist paint effect, such as ragging or sponging, or even a fake marble finish may be authentic. When you view the final effect, you will see that such finishing touches are as important as the features themselves.

A fine fireplace with ornamental plasterwork like this is always worth restoring to former glory.

INTRODUCTION

STYLE GUIDE

*W*herever we have used stone or rock to build our homes and temples, it has endured, whether as barely discernible ruins, roofless huts or complete and habitable buildings for us to study and marvel at. Many of the more important structures have been restored and improved – even updated, over the centuries: the houses of the wealthy and the churches and temples of the major religions. Where war or decay have intervened, the stone has often been purloined by the local inhabitants to be recycled into new buildings. Many a home (or church) has identifiable parts from some Roman garrison town that once stood on the site or from a nearby castle, now in ruins.

You do not have to be a student of architecture to have your imagination fired by the changing building styles and techniques that can be discerned to have evolved through the centuries; observing both interior and exterior shape, scale and finish of a dwelling can be fascinating. It is not just the basic shell of a building that gives it character, but also the added features – the porticos and porches, doorways, balustrades, balconies, steps and staircases, garden walls and even flagstone paving, whether inside or out – that have been designed to complement or transform the main building. These may be overlaid with further ornamentation in stone or plaster: stone pine cones (often mistaken for pineapples), figures, animals and heraldic devices relevant to the era are frequently seen. The Tudors, for example, used terracotta to create animals, flowers and coats of arms, while the

STYLE GUIDE

Jacobeans favoured plasterwork ornamentation. Classical swags, pillars and porticos indicate the Restoration period. Eighteenth-century Romanticism incorporated niches for displaying figurines, urns and cherubs. The seventeenth and nineteenth centuries had a passion for pargetting (moulded plasterwork) and stucco (smooth or modelled plaster). Then again, there are the many variations of materials and techniques identifiable from region to region according to locally available stone and skills.

It is when you buy a period property of your own, however, that your appetite for detail and historical information will really be whetted, especially if there is restoration or replacement work to be done. Whether the necessary work is inside or out, it is always worth researching the subject and making a proper job of it, using, as far as is practical, the correct materials and the right techniques, and finding the closest possible match in style. There are specialist books, old catalogues and style guides, as well as restoration and reproduction companies that can advise you. One of the best ways to conduct your research is to study other buildings from the same period or by the same school of architects, not just in museums and on historic sites; neighbouring buildings can often offer clues and information or fill in the gaps where features may be damaged or missing. Of course, if your property is of special historic interest, is listed, or lies within an historically designated area, you will be compelled by law to take this kind of care, and the work will be inspected. The plus side to these restrictions is that grants are sometimes available for the work, as is professional advice, so it is always worth checking out the situation with your local authorities before any work starts.

Take a single feature as an example: fireplaces are one of the most popular restoration projects for enthusiastic home renovators. There was a time when these were all ripped out in favour of central heating, leaving many rooms architecturally diminished and lacking a focal point. Now everyone is keen to put them back in, and original stone and plaster surrounds, once to be picked up free off a skip, are

fetching high prices. In your own home, it is worth checking whether a much older (and more relevant) fireplace lies behind the existing one; sometimes it is worth removing an ugly or inappropriate later model in favour of the original.

When restoring or replacing, try to take a look in neighbouring properties or in similar buildings from the same period, remembering that scale and style are as important as historical accuracy. The type and size of fireplace used to reflect the importance and scale of the room, becoming more modest as you approached the bedrooms and the smaller, less frequently seen rooms at the top of the house. Replacement fireplaces can be found in specialist restoration yards, but be prepared to spend some time finding exactly the right one. Some suppliers will actually bring a selection of suitable models to your home so that you can see how they look *in situ*. Be very sure, when buying a reproduction fireplace, that it is a good one and that the manufacturers have got the proportion as well as the style and materials correct; often decorative features will have been scaled down, losing impact and credibility. If yours is a Victorian property, or even a century older, you will find plenty of useful references in old catalogues and existing period models.

Sometimes the fireplace has not been removed, but simply blocked in, and uncovering an intact feature can be a real treat. Always check that the flues and chimney are in proper working order, or indeed have not been removed altogether, before starting work on the actual fireplace; a chimney specialist will be able to advise you on whether it needs cleaning or lining to be safe and efficient in operation. Sometimes a fireplace has simply been painted over (page 34), or maybe parts of it are damaged or chipped, or there are even whole parts missing. A broken lintel, chimney bar or brick arch is common, and these may need temporary support during repair for safety. Always approach the project carefully in order to avoid further damage.

The earliest style of fireplace (beyond an open fire escaping through a hole in the roof) was a simple hood set in the wall over the fire. This

was supported by piers or a stone lintel bracket, a feature that survived for centuries in more modest homes. Then richer homeowners began to embellish the structure, developing the fireplace almost as an art form and incorporating its design in the overall decoration of the room. By Tudor times, it had been recessed into the wall as part of the general panelling, with a fancy stone arch; and for the Elizabethans it had become an important focal point, garnished with elaborate columns.

The seventeenth century even saw imported styles being brought over from Europe – mainly Italy – and it was at about this time that clay bread ovens were incorporated into the fireplace, a feature still seen today in many country homes and one well worth restoring. By the eighteenth century, even the more modest houses were attempting to imitate the grander homes and pattern books were published; fireplace designs in the classical style by Inigo Jones were particularly popular. Often, the fireplace would be designed to imitate the style of the door-frame, using columns or pilasters and an entablature supported by consoles, with a picture panel above to set the whole thing off. At first, these were constructed in timber, but soon stone or marble was used; white marble was the preferred material.

Style became increasingly elaborate and refined; by the late eighteenth century, fine ceramic cameos were being incorporated, many produced by Josiah Wedgwood. Georgian style became typified by delicate fluting, as a lighter, less heavily decorated look was preferred, a genre perfected by the Adam brothers, who employed delicate classical motifs. Marble and stone had become expensive by the mid-nineteenth century, and often cast-iron fireplaces in minor rooms would be painted to look like stone. For more important settings, such as drawing rooms, the Victorians and Edwardians could choose from mass-produced marble surrounds, with the traditional picture or stucco panel above replaced by an impressive mirror (later to be developed into a full-blown overmantel).

The fireplace was to be influenced by all the current fashion trends

in interior decoration (Gothic, Arts and Crafts, Art Deco), but, as marble became expensive, they became smaller, and slate became the preferred material. Tiles were often inserted into the structure, and these also reflected popular fashion; they are often damaged or missing but are easily replaced, either with originals or faithful reproductions, often produced by the original manufacturers, using traditional methods, from the original moulds.

Fireplaces are only one fascinating architectural stone and plaster feature worthy of study and restoration. We were building our homes in stone at least 5,000 years ago when Neolithic people erected their huts in Britain, although these techniques were forgotten after the Romans left, and not revived until the Normans invaded. With the Normans came the start of 500 years of impressive stonework, with tools and techniques improving and developing over the centuries. Many fine castles, churches and manor houses survive today as testament to the quality of the craft. By the fourteenth century, the skills were clearly defined, and a kind of hierarchy had developed among the craftsmen. Top of the tree was the master mason (he was often the architect too). Under him were the freemasons, who tackled the more specialist work, such as carving. The rough mason's job was to cut and dress the stone, while labourers fetched and carried it.

Stone always had a reputation for representing quality and wealth, so the earlier houses built of solid stone (that is, before the sixteenth and seventeenth centuries) tended to be the grander ones, except where stone was readily available from a nearby shallow quarry or demolished building. However, by the seventeenth century, smaller stone houses were being built by skilled masons, particularly in limestone areas such as the Cotswolds, where good building stone was plentiful. Style varied according to availability and budget. Rubble (irregularly shaped and sized stone) might be laid randomly with mortar joints between, or laid as closely as possible in courses, often with brick or shaped stone quoins (corners) to produce a more regular appearance. Properly squared-up stone – called 'fine ashlar' – was expensive, and could be

laid with neat, narrow joints. This was reserved for quality buildings. A compromise was 'rough ashlar' – a roughly squared-up stone – which was cheaper and could be laid in regular courses, although with wider joints.

Other regions also began to employ whatever stone was locally available for extensive building: flints were readily available on chalk downlands but were rather irregular, so tended to be mixed with more regular stone or even brick. This would be used to build the quoins, or might even be laid in bands through the more irregular stone. Not only was this decorative, but it also added strength to the construction. Some areas such as East Anglia had the flints knapped and cut into square shapes which could be used to create ornamental patterns when mixed with stone. In the late eighteenth century, an artificial stone was developed. Called Coade stone, it was used for mass-produced features until the 1830s, and has proved surprisingly resilient. Sometimes features were made in plaster or cement and painted to look like stone, in an attempt to economize.

Although stone walls were thick, in order to keep out the cold and damp (page 22), plastering was introduced as a way of further reducing heat loss and of improving resistance to fire, and, in the case of wattle-and-daub construction, as an attempt to stabilize the walls. Inside the building, bare stone, as is the fashion today in houses of certain periods, was rarely seen. Unless constructed in smooth-faced, regular ashlar stone, walls would be plastered. Sometimes the plaster was marked into sections in imitation of ashlar stone. Timber-frame buildings were also plastered within, then painted to look like timber framing. In medieval times, the plaster was simply a mixture of lime and sand, applied thinly. The seventeenth-century plasterers reinforced the mix with goat or ox hair and applied it in a thicker coat; by the late eighteenth century, a harder plaster was available, thanks to hard hydraulic lime (Roman cement) and Portland cement, which could be applied with a metal trowel to produce a smoother finish very like today's gypsum-based plasters.

Like other practical techniques with simple origins, this gradually evolved into an art form, typified by the Tudors whose plaster ceilings imitated the style of earlier decorative timber forms, with sections divided by moulded ribs, plaster bosses or pendants embellishing the intersections in the style of stone vaulting. Henry VIII's palace at Nonesuch was a good early example. Gradually the ribs became flattened to become the familiar strapwork design decorated with lozenges, scrolls and roses. These could be made in moulds, some of the more detailed figures being modelled on site. A decorative plaster frieze was sometimes applied where the wall met the ceiling, as a kind of decorative coving, and a wide range of elaborate motifs were available. Poorer homes plastered over their beams and joists and applied more modest ornamental panels.

The Tudors and Jacobeans modelled plaster into panels and friezes showing mythological hunting scenes, complete with plants and animals. By the 1620s and 1630s, Inigo Jones had made smooth plastered and painted ceilings fashionable, incorporating covings, mouldings and gilded details which might include fruit, flowers, wreaths, ribbons or cherubs. Later, dolphins and more stylized decoration became popular themes. This classical style was popular right into the eighteenth century when hand-crafted plasterwork was also loved by the Palladian architects, who designed decorative ceilings, cornices and friezes as well as ornamental architraves, frames and panels in the French style of the period.

By Regency times, tastes were more refined, and a highly ornamental style was considered rather heavy and ugly. The Adam brothers put their family name to a particularly distinctive style of finer plasterwork, which was ultimately painted in pink, green or blue pastel colours. The Adams mixed a special kind of plaster containing gypsum or fibre and

Dressed ashlar stone has always been expensive and is traditionally reserved for grander, more formal buildings.

glue (called Liardet's Preparation) which was not modelled on site but pressed into hot metal moulds according to careful measurements. These later inspired the mass-produced mouldings of the 1800s which still influence room design today.

The Georgians enjoyed relief plasterwork, and even fairly modest homes might be quite lavishly decorated with classical figures, birds or the ubiquitous acanthus leaves. Plaster panels imitating current Italian trends became fashionable; these were used to cover beams and update interior decoration. Poorer homes featured a simple cornice and plain plastered ceilings.

As the nineteenth century moved into the twentieth, the well-off homeowners returned to richly decorated themes such as landscapes and mythical figures with classical columns in stone and marble. By now, mass-production had widened availability to many more homes. Cornices and friezes could be moulded; sometimes papier mâché was used as an alternative for ornamental work, as it was light and easy to fit with screws. Mass-produced items also permitted repeated themes, although this naturally lacked the originality of the early hand-crafted work. Ornate ceiling roses and heavy cornices began to be seen everywhere in a great many styles and types. Years of overpainting will often have reduced detail and diminished their impact, so they will need careful cleaning (page 39).

There was a revival in craftsmanship at the end of the nineteenth century, with decorative features carefully geared to the shape, size and function of each individual room. Popular themes were oak trees, acorns, flowers and animals. However, this coincided with a boom in mass-produced plasterwork, using clay models to make jelly moulds, which were extremely popular although they could not offer fine detail. Some of these moulds survive today and are useful for manufacturing replacement plasterwork.

Plasterwork, whether plainly functional or highly decorative, was not restricted to the interior rooms of a building. An external coat of mortar, called render, has been used to protect and embellish the

outside since Roman times, although gypsum-based renders were not used until the mid-thirteenth century, and artificial cements not introduced until the late eighteenth or early nineteenth centuries. As with the internal plasterwork, different styles have gone in and out of fashion through the centuries. From 1784 to 1850, for example, there was a tax on bricks, so external plasterwork and, in particular, stucco – a rainproof, close-grained, hard plaster finish, which could also be painted – was very popular, and was also used as a way of updating the appearance of earlier timber-framed buildings.

Stucco, as a form of modelled plasterwork, became fashionable in the late sixteenth and early seventeenth centuries in imitation of current European (especially Italian) styles. Both stucco and other sand/cement renders, such as pebbledash and roughcast (page 37), were especially popular in Regency and Victorian times for providing weather-protection to porous, thin brick walls, which saved on construction costs. Actually, although render will resist windblown rain, once the wall has become saturated the damp is drawn through; this dries out afterwards, and can be seen in colour variations on the wall according to the weather. The Victorians favoured cornices and mouldings on their stucco, and these were often prefabricated in sand/cement render on a bench on site. The fashion for stucco had died out towards the 1870s, and pebbledash and roughcast renders became popular towards the end of the nineteenth century.

Regional fashions can also be found in external plasterwork, such as the decorative pargetting most often seen in East Anglia, but also found in other parts of England. This involved a mixture of lime, sand, animal hair and animal dung being applied to both external and internal walls in ornamental patterns, such as herringbone, basketweave and even animals, birds and figures, to cover walls, overmantels and gable ends. This fashion died out by the end of the nineteenth century, owing to a lack of suitable-quality plaster.

In order to keep your restoration work in period style, always follow the guidelines on page 20.

*Above: Rough stone mixed with brick is often a
regional architectural style.
Right: Fabulously ornate: a restoration job like this will require
professional advice.*

STYLE GUIDE

STYLE POINTERS

- Never remove original details, such as plaster cornices, friezes, mouldings etc., or even an old plaster ceiling, unless strictly dangerous. Repair, wherever possible, using traditional techniques and employing reclaimed parts or new pieces moulded on original features.
- Do not expose large areas of stonework. This was usually plastered. Resist the temptation to leave the odd stone sticking out to 'add character'.
- Be careful to apply a rough finish to plaster if reproducing an era when modern tools were not available. Avoid sharp, geometrically correct corners in older properties (never use mesh reinforcing corners). Blunt the edges slightly to imitate centuries' wear and incorporate a gentle, almost curvy profile, without making it look too 'wonky'! In simple cottages, use a cross-grained wooden float on new plaster, while still wet, to texture it. Follow the natural contours of the wall using a weak mix of 1 part cement, 2 parts lime and 9 parts coarse sand.
- Never seal interior stonework with varnish or a similar sealant that prevents the wall from breathing naturally.
- Always repair stonework with the correct size and type of stone to match the original, taking care to lay the stone facing the same direction and mortaring with a matching type and colour. Never be tempted to speed up the job by mortaring larger gaps than necessary between the stones.
- Do not change the proportions of a period room by lowering a ceiling or partitioning it off so that a fireplace looks too large or appears off-centre, or where decorative mouldings or other round-the-room details are divided. You may have to add new sections to make a newly divided room look correctly proportioned.
- Never pick out mouldings or other plasterwork details in coloured paint or gilding without first checking the correct period style.

TOOLS AND TECHNIQUES

The tools and techniques for masonry and plasterwork have changed very little over the many centuries they have been employed. The workers who built the ancient pyramids of Egypt would have easily recognized the hammer, chisel and plumb-bob used by today's masons; the only relatively new addition to the stoneworker's toolkit is the spirit level – invented in the seventeenth century. If you are considering the restoration of a stone structure or plasterwork feature you will have the satisfaction not just of preserving a little piece of history but also of continuing a very long tradition of craftwork.

The damage you are most likely to encounter externally is weathering. Stone seems very durable but, although it does harden and form a kind of protective skin over the years, it is in the most part porous, which makes it susceptible to the ravaging effects of water, frost and heat, and, increasingly of pollution such as smoke and chemical emissions. In a coastal site, a salty atmosphere can also be devastating if no regular reparation work is carried out over the years. Pollution is today's greatest enemy to old buildings: the chemical constituents of stone can react with certain chemicals in the atmosphere and start to break down. This process was accelerated in industrial areas, such as cities, during the Victorian era, when the atmosphere was saturated with polluting substances created by domestic and industrial coal fires.

Section Two

One of the constituent parts of soot is sulphuric acid, which reacts particularly badly with limestone and calcareous sandstone (two of the most popular building stones). This changes the normally insoluble calcium carbonate into calcium sulphate, which is soluble, and thus wears away the stone. Rain containing dissolved soot is heavy with carbonic acid, which also reacts with calcium carbonate, turning it into soluble calcium bicarbonate. Today, emissions from motor vehicles are the culprits, eating away our stonework.

Sometimes the very construction can cause stone to corrode or disintegrate: for example, it is important that the blocks are laid correctly – that is, with the bedding lines running horizontally, even on vertical structures such as columns. Laid the other way (called face or cant bedding), the building will eventually have to be demolished. Similar problems will be encountered should limestone be laid next to sandstone, as they react together and start to disintegrate. Occasionally, you do find different blocks of stone banded together, or even sandstone rubble within limestone columns. As water runs from one to the other, a chemical reaction takes place and the salts crystallize. A more common problem is the application of too hard a mortar; the mortar should always be less hard than the stone, so that the moisture entering the wall evaporates through the mortar and not the stone. Beware also of those iron cramps you frequently see intended to hold the stones together – many of them are unnecessary, incidentally. As the cramps rust, they naturally expand and split the stone. Where they might be structurally essential, they can be replaced (using X-rays or a mine-detector) with a non-corrosive material such as copper or stainless steel.

Dressed stone has always been expensive and is not always appropriate for less formal, humbler dwellings. You will find many cottages and larger farmhouses built in previous centuries still solidly built of rubble (stones of random size and shape). However, the walls must be at least 450 mm/18 in thick to be effective against cold and damp, with joints no bigger than 13 mm/½ in thick, which can be

This stonework has been seriously damaged – probably by movement in the building.

expensive and time-consuming when using small stones. The structure must also have good-quality bond ('through') stones running at least two thirds of the way through the thickness of the walls. In the better type of random stone building, the individual stones are arranged more or less in horizontal courses, or a more random arrangement might have been used, with brick or dressed stone quoins (corner sections) to give a sense of formality.

MATERIALS

STONE

Stone is generally quarried (although a little is mined). In the old days, plugs were hammered into the rock, then soaked with water, so that they swelled and split the stone (today, compressed air drills are used

Splitting blocks of stone in the quarry by drilling holes along a line of weakness and inserting plugs.

to make the holes along a line of natural weakness in the rock). The holes were fitted with wedge-shaped iron pins, which were sledgehammered. This method produced large blocks or quarried stone that subsequently had to be sawn into smaller, more usable pieces by a stone sawyer, whose modern equivalent is the mechanical saw. These are available in various sizes with tungsten or carborundum toughened blades, kept cool with running water.

Cutting the harder types of stone by hand used to be a tedious business, even with a double-handled saw. It was tough on the saw, too, although the procedure was later helped by the invention of the frame saw. This could be worked by one person, and had a long blade suspended by a pulley to take the weight of the stone, cutting in a vertical action using sand and water. The stone sawyers cut the stone into a variety of roughly standard shapes, producing a rectangular face. A few were cut more carefully, with proper angles and a smooth face, for stone that is destined to be on show, set in a horizontal course. These were called ashlar blocks, and were always expensive.

Further skills were employed on the stone by the 'banker masons', whose job it was to dress the outer surfaces of the blocks; they would also undertake any carved shaping. As they finished each block, it would be numbered for positioning on site. The procedure was to make a template of the required design in cardboard or zinc and then trace it onto the stone in pencil, having first checked the block for size and angles with a set-square. The stone was roughly chipped away around the design, using a wooden mason's mallet (to absorb the shock) and a broad-bladed chisel with a blunt edge. The mallet was usually made from a strong dense wood, such as beechwood, to resist splitting. The mallet had a cylindrical head, tapering slightly towards the handle. The finer chipping away to complete the design was done using a 'waster', which takes away a thin shaving of stone and is more controllable. To finish off, the masons used a 'boaster' – a type of sharpened chisel which could be used to make an accurate straight surface. Rounded shapes and hollows could be cut with a gouge similar

to that used in woodworking. Awkward shapes could be cut using a handsaw. For the finish, there was a choice of many techniques producing different textures and patterns, as listed below.

- 'Drags' are semi-circular tools with a toothed straight edge which have different-sized teeth to produce a variety of finishes, from a criss-cross pattern to rough lines. These indented lines increase weathering on the stone – a deliberate intention of the process.
- For a fine smooth finish, the surface was rubbed with a piece of harder stone lubricated with sand and water. Where the stone itself was exceptionally hard, such as granite or marble, the surface could be polished with turpentine or beeswax and a soft cloth.
- The surface could be 'scabbled'. This involved taking out scoops of stone, using a heavy, pointed hammer, creating the kind of finish achieved on timber with an adze. For a snug fit, blocks with this type of surface decoration were often finished with a narrow border called a 'chisel draughted margin'.
- Sometimes the boaster was used to cut parallel marks across the surface of the stone; this was called 'droved work'. If the lines were

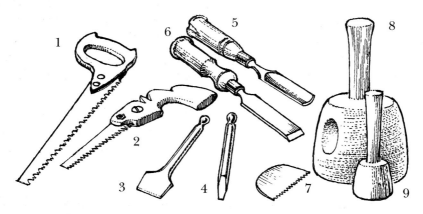

Masonry tools: 1 mason's saw; 2 adjustable blade saw; 3 claw boaster; 4 waster; 5 gouge; 6 chisel; 7 drag; 8 mallet; 9 dummy.

continuous, they were called 'tooling'; deeper parallel lines like grooves were called 'furrowing'.
- For a grainy effect, which made the stone look smooth from a distance, a pointed punch was used all over the face.

Further carving of the stone might be undertaken, where the most intricate designs were required, once the block was in position, by a sculptural mason whose skill enabled the fashioning of lifelike plants, animals and figures, or heraldic designs. Finally, the blocks would be put into position on site by the fixer masons, who would mortar and trim the stone in place.

Masonry techniques do depend very much on the type of stone being used: soft or hard, durable or crumbly. Its resistance to weathering will also affect how and where it will be used, especially when considering the load and stress of a large building. The builder/architect will have had to take into account all these practical considerations as well as the colour and texture of the stone. Until the industrial revolution in the nineteenth century and the development of a more efficient means of transporting heavy loads, the mason would be restricted to what could be obtained locally, although there are improbable instances of stone being transported long distances for special projects – Stonehenge, for example. In Norman architecture, there is also evidence of stone being imported from France. Generally, though, there are three main types of rock used in building, as follows.

Igneous (for example granite): once molten rock, this type of stone is very hard, which makes it very difficult to cut but extremely durable. It can be ground and polished. Cutting it wears out tools quickly, although today you can use tungsten-tipped chisels.

Sedimentary (for example limestone and sandstone): these rocks are made mostly of tiny particles, and because they can be worked easily in any direction (and are thus called freestones) have always been very

popular for building. This group covers a wide range of types, varying in hardness and durability. They come in a wide choice of colours, too: limestone can be white through to every shade of brown; sandstone can be brown, cream, pink or even red.

Metamorphic (for example slate and marble): these are igneous or sedimentary rocks that have later been altered by heat and pressure. Although quite fragile, slate is a popular building material, frequently used for door and window lintels or floor flags, as well as split into thin sheets for use as roofing tiles. Marble tends to be used more in locations with a Mediterranean climate. It can be solid, or split into tiles to be used on walls and floors. Its austere, almost clinical, expensive look is better suited to hot-weather homes, where its natural coolness is ideal.

MORTAR

Mortar is used to bed in the stones in the initial construction process, then later the joints between the stones are filled with mortar or 'pointed' for a neat appearance and to improve weather resistance. Since this surface mortar tends to crumble and fall out after a time, repointing is one of the most common and regular repair jobs necessary with stone walls. At the outset it may seem a simple (albeit boring) task, but if done badly it can ruin the whole appearance of the building. It should never be applied proud of the wall or left smeared over the surface of the bricks (called 'buttering'). The most important consideration is the colour of the pointing, which should be matched as closely as possible to the original, by careful choice of ingredients or the addition of either stone dust or a little of the old mortar, crushed. Mortar is basically sand (the filler) mixed with lime or cement (the binder). Today, cement is used almost universally, unless the mortar is being used in the restoration of old buildings, but this has only been in widespread use since the beginning of the twentieth century. Although a little cement in the mix can help the hardening of hydraulic lime

mortars (see below), a lime mortar is preferable when working on older structures. Portland cement, for example, sets very hard, which may not only crack the stones but also impede the natural 'breathing' of the wall. It also shrinks on setting, and, unlike lime mortar which is a kind of creamy off-white, is mostly grey, which can look obtrusive and ugly.

Concrete: a durable stone-like material comprising coarse and fine aggregates (particles of natural stone or sometimes manmade materials) bound by hardened Portland cement paste.

Portland cement: available in light grey or white, Portland cement reacts with water to form a dense hard mass, virtually insoluble in water and unaffected by most substances except acids and some chemicals.

Mortar: the term generally means a mix intended for bedding masonry or paving or for finishing masonry (pointing). In fact, it is a form of concrete, containing only fine aggregates.

Masonry cement: a type of cement made for use only in mortar for masonry or rendering, not concrete or floor screeds.

Aggregates: these generally comprise naturally occurring sand and gravel, or crushed stone usually quarried locally, and classified as coarse or fine, depending on whether they pass through a 5-mm/$\frac{1}{5}$-in sieve. Ordinary concretes contain a mixture of both, with a maximum size specified for the coarse material – usually 20 mm/$\frac{4}{5}$ in, but sometimes 10 mm/$\frac{2}{5}$ in or 40 mm/$1\frac{3}{5}$ in. Specially selected alternative materials may be used to produce a particular colour or finish. The sand can be 'sharp', which is the coarsest type for use in cement, or 'builder's' (soft) sand intended for masonry mortar.

Chemical admixtures and pigments: there are various additives which can be used to improve certain properties of cement and mortar –

waterproofing and frostproofing, for example. Limes are generally identified by type as being hydraulic, non-hydraulic or semi-hydraulic. This refers to their suitability for setting under water. Hydraulic must be used quickly (within around four hours); non-hydraulic can be kept soft indefinitely. The three types vary in colour, too, from the grey/brown hydraulic to the white non-hydraulic, which is purer in limestone. Choice of lime generally depends on local availability and is produced by burning limestone in kilns to make quicklime (calcium oxide). This oxide must then be hydrated by water (known as 'slaked') to form lime putty (calcium hydroxide) – a rather risky process, producing much heat and dangerous gas. The putty will last for years. Today, you can buy ready-hydrated lime in powder form to be mixed with water into a putty which is usable after 24 hours and much safer. You can buy lime putty in bags – this is not 'bag lime' available from builders' merchants. The setting time of the lime is controlled by the addition of brick dust, wood or coal dust – or, these days, pulverized fuel ash or cement. The ingredients for mortar are always carefully measured to a set ratio: cement : lime : sand for non-hydraulic and semi-hydraulic limes; or just lime : sand if cement is not being used. For example, a popular lime mortar mix is 1:1:6. You can buy sand and lime mortars ready mixed from DIY stores, but they tend to make too hard a mortar; if the mortar is stronger than the stone, you will get moisture problems. The ingredients are usually measured by weight or volume and mixed carefully with water to the right flexible consistency.

REPOINTING

Old mortar may be crumbly from frost or from bad preparation, but there is usually no need to repoint unless the stones are working loose or water is penetrating the wall. Repointing poor or damaged sections tends to produce an unsightly patchy effect, so, if you are keen to have a perfect finish, it will have to be the whole job or nothing. This

Cleaning out the old mortar and other debris before restoration.

involves raking out all the old pointing with a flat-bladed, pointed plugging chisel, called a 'quirk', until you reach the hard bedding mortar. The joints should then be washed out with clean water and the new lime mortar forced into the gaps with traditional metal tools. Repointing should never be tackled in frosty weather.

Start at the top and work downwards, making sure you match the finish to the original: the most common type of joint in period buildings is flush to the surface of the wall, although you sometimes see slightly recessed pointing, known as 'hungry' or 'hollow key'. It used to be created by drawing the handle of a bucket across the face. Never let the pointing stand proud of the stone (this is called 'strap' or 'ribbon' pointing – it is not only ugly but also does not do its job properly, as the water does not run away as it should).

Lightly brushing the mortar after it has set prevents too smooth a finish. Avoid smearing the face of the stone with mortar as much as possible; this is called 'buttering' and, where it has occurred, you should brush it off meticulously the day after repointing. Some areas have certain decorative traditions such as painting or scoring the

pointing, and these should be followed where appropriate. The fine pointing between ashlar blocks should only be tackled after professional advice.

CLEANING STONE

A stone surface will often require cleaning before any repair or restoration work can begin. In a country location, where buildings are more expected to mellow and blend with their surroundings, this might be less important than, say, in a town where a smart appearance traditionally denoted affluence and good character. You may even require planning permission before cleaning a property of particular historic importance. Different types of stone will require different treatments, and if you have any doubts you are advised to consult a professional cleaner.

In all cases, testing a small, little-seen patch before tackling major areas is advisable. An expert will take care to plug any cracks and open joints, and to protect windows, before starting any cleaning work. One method is to spray with cold water and scrub gently with brushes and wooden scrapers; this is particularly suitable for limestones and marbles. It is important not to soak the wall and to check with the local water authority that there are no damaging salts present in the tap water. Never attempt cleaning when the weather is frosty.

Dry-grit blasting will clean hard stone such as granite and slate (providing they have not been polished) but is no longer as popular as it was for softer sandstones. There are disadvantages: operation of the equipment can be dangerous and is tremendously noisy, and there is a risk of drains becoming blocked. In addition, the immediate surface of the stone is actually removed in the cleaning process, which may not be acceptable. Equal caution is required when considering chemical cleaning. This may be suitable for some sandstones and unpolished granites, and can actually be effective against lichen, mosses and graffiti. Beware, however, of using acids on limestone.

Of course, you can simply scrub the dirt off: power tools are not really to be recommended on period properties, especially if they are already in some state of decay, so hand-held brushes and scrapers will be the order of the day. Remember to go gently on fine areas and not to wet the surface too much. Particular stains can be removed as required; green stains are usually due to copper carbonate produced by rain running onto stone from brass, copper or bronze. This can be treated with an ammonia solution and a special poultice. Salt thrown up from winter snow-clearing will sometimes leave a deposit on stone surfaces, and this should be dampened and wiped away as soon as possible.

STONE RESTORATION

Damaged or missing sections of stone will have to be matched: normally, the stone will be of a local type, and the local quarry may be able to help, if still operating. There is also a brisk trade in stone reclaimed from other buildings, which may be reworked or retrimmed

Stone can be trimmed roughly to fit using a hammer and boaster.

to fit. Any delicate carving or detailed work that has to be matched will be a skilled job and it is usually best to consult the appropriate craftsperson. Existing sections and neighbouring buildings may provide the clues for missing parts; old architectural pattern books and guides may also be of assistance. A camera and notebook and a pair of mason's callipers for measuring the exact dimension of a feature can all be useful additional tools when researching authentic features.

RESTORING A MARBLE FIREPLACE

Although less prone to weathering, internal stone features may also be in great need of cleaning or repair and, again, attention to historical detail and to the correct processes is essential. An existing marble fireplace is surely worth renovating to former glory, but it may be stained, chipped or even painted over. The project needs approaching with caution, for, despite the fact that marble appears hard and durable, it is actually very susceptible to being ruined by acidic or alcoholic cleaning preparations: the stone is in fact porous and corrodes easily.

Paint: paint can be stripped off successfully using a proprietary paint-stripper, but you must take suitable precautions, such as wearing goggles and rubber gloves, for your own personal safety. Marble is easily scratched, so remove the paint with swabs of cotton wool dampened with acetone rather than a scraper.

Staining: staining is a frequent problem with marble fireplaces. You can buy from fireplace and marble specialists a proprietary cleaner, such as Sepiolite, which you mix to a paste with de-ionized water and apply to a thickness of about 1 cm/⅖ in. This is left for about 12 hours, or until it begins to dry out and cracks start to show. The paste should be removed with de-ionized water on cotton-wool swabs.

Chipping: you can buy marble glue in a choice of colours to match various marble types, and these are simply applied to the damaged area. Once they have been allowed to set hard, rub them into the desired shape or profile using fine-grade wet-and-dry sandpaper.

Worse damage: if larger chunks and chips are missing, you can buy marble offcuts, which can be cut roughly to shape and applied using marble glue, then rubbed to a finer, more accurate finish as described above. Damaged carving or similar detail will require professional help.

Polishing: it is important to use only a white polish, as a coloured one will stain the marble. Rub it in hard, using a soft cloth; again, it is important not to use a coloured cloth or duster because of the risk of staining. Keep the marble buffed up once restored.

RENDERING

External render was initially applied as a means of weather protection, which is why you often see it in isolated patches – on a north-facing wall or on chimneys, for example. Later, it was to double as a decorative finish, from simple scoring (to imitate more expensive dressed stone) to textured and moulded decoration. Render is traditionally mixed from lime, gypsum and cement. The Romans used lime and natural cements (Roman cement is the best-known hydraulic type); gypsum renders have been used since the thirteenth century; artificial cements like Portland cement were not introduced until the late eighteenth or early nineteenth centuries. The cement renders tend to be a little too strong for period buildings, and, although Portland cement is used extensively, lime render is far superior in that it allows the wall some movement, lets the stone breathe and has a softer, more mellow appearance. The minor disadvantages are that it is slow to apply and is likely to shrink and crack. The addition of gypsum

(a natural product burnt in a kiln, like lime) will strengthen a lime render and improve its setting qualities, but has a tendency to go too hard, like a cement render. It may also crumble in damp conditions.

The surface to be rendered must first be prepared by roughing it up to provide good adhesion for the first render coat; this is called 'keying the surface'. It involves raking out all the mortar joints and scoring the surface with a nail knocked through a wooden float. This is particularly important if the wall has been distempered or coated in a water-repellent finish. What is essential is that the surface should be clean (page 32) and damp, but the weather never frosty. Three render coats are usual, especially if the site is exposed, but in some early buildings a single 'daub' coat has been sufficient. Two days in summer and seven in winter are generally allowed between coats, then the surface is scored or scratched again to provide a good key. The first coat is called the render coat, the second the floating or 'backing' coat and the third

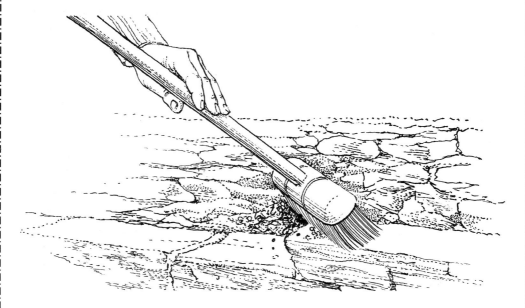

All dust and debris must be brushed away from the surface of the stones.

the setting, finishing or smoother 'skim' coat. Each coat should be thinner and contain less cement than the one before. The final skim coat is traditionally finished with a wooden float, which produces a flat but slightly textured finish that weathers well and is more interesting than the effect created by a more modern steel float. There is also a range of different finishes, introduced over the centuries or prevalent in certain regions, for adding texture or decorative interest, as follows.

Stucco: the word is Italian and is used to describe a superior-grade plaster, introduced in Italy by the late nineteenth century, and used for all kinds of external plasterwork on buildings requiring a smooth, expensive-looking finish. It was often moulded to imitate formal stonework. By the end of the nineteenth century, the standard recipe was recognized as being one part Portland cement to three parts clean sharp sand. A little lime in the mortar improved elasticity, and pigments such as Venetian red and yellow ochre were added to create coloured finishes. The Victorians also favoured elaborate stucco mouldings and cornices cemented onto the stucco with render.

Roughcast: a roughcast finish takes at least two coats: the initial 'straightening' coat is a 1:1:6 cement/lime/sand mix plus animal hair to bind it. This is combed to provide a good key, then left to dry. The second coat is a wet mix of 1½ cement/½ lime/2 shingle/3 sand flung against the wall, using a hand-scoop or laying-on trowel, to produce a rough-textured, pebbled finish. Roughcast is usually painted, as its natural colour is a rather unappealing grey. There was a fashion in the sixteenth century for including fragments of glass in the aggregate, which produced a sparkling effect. Sometimes, also, you will see quirky houses (in coastal areas, for example) where coloured pebbles or shells have been painstakingly pressed into the damp render to create pictures and patterns.

Pebbledash: an alternative pebbled finish can be achieved by throwing

coarse aggregate, such as spar or pea shingle, onto the render with a hand scoop while still wet.

Pargetting: this decorative form of external plasterwork, mainly found in East Anglia, displays intricate relief patterns in plaster, created in beeswax moulds or cut, using combs. They were originally painted in a range of colours, but few painted examples survive. More often, the designs have been repeatedly limewashed and lost some definition. Repairs must always be carried out using a lime (not a cement) mortar, using a 1:6:9 mix and employing the same proportions for both undercoats and topcoat. Some animal hair in the undercoat mixes can be useful; this will increase strength.

INTERNAL PLASTERING

Before the development of plasterboard, plaster had to be applied to a framework of horizontally slatted wooden laths, nailed at close intervals (approximately 6–10 mm/¼–⅖ in apart) across the timber

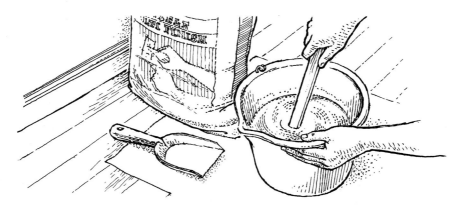

For a small patching job, or if you want to practise your technique, mix up just a bucket of plaster or render.

framing. The thin softwood laths generally measured around 25 mm/ 1 in wide by 6 mm/¼ in thick by 1.2 m/4 ft long, and the plaster comprised lime and sand with the addition of horse or ox hair; you can usually patch a damaged area with modern gypsum plaster, such as Carlite, which will be compatible with traditional lime plaster and should adhere well to the laths. If the laths are damaged, you will have to provide an alternative for the plaster to adhere to. A small hole no more than 75 mm/3 in wide can be breached with a piece of scrim (jute sacking) soaked in plaster of Paris; larger holes will require sections of plasterboard or expanded metal lathing (first introduced as a substitute for wooden laths in the nineteenth century) nailed to the studs. The room may require restoration of more decorative plaster features; fibrous plaster mouldings prefabricated from a mould in a factory and erected on site have been popular since the mid-eighteenth century, and there is a wide choice of fibrous plaster rosettes, medallions and other features still available today and taken from existing originals. However, where *in-situ* runs of decorative plaster, such as cornices, have been damaged, it is advisable to seek the help of a professional. This can work out to be very expensive, as there is a lot of remoulding involved. Renovation of decorative plasterwork should only be carried out once the flat bed of the ceiling is clean and sound (see below).

CLEANING PLASTER

Sometimes centuries of paint or distemper can virtually obscure all the detail on an elaborate cornice, ceiling rose or other ornamental plaster device. You can strip it off by hand (very carefully) using a proprietary chemical stripper, but this can damage the plaster and it will need to be repaired immediately after stripping. Patience and care are paramount: to remove water-based distemper, moisten the feature with wet cloths or sponges, although a steam-operated wallpaper-stripping

machine is better (these can be hired). When damp, gently scrub the relief, using a small brush or tools like those used by woodcarvers for awkward crevices. Never be tempted simply to paint over distemper: the paint either flakes off or seals on the distemper permanently. To remove old emulsion paint, try steam-stripping or methylated spirits.

REPAIRING A CEILING

It is vital to check out the soundness and to make the necessary repairs before resculpting or redecorating a ceiling. Look for the tell-tale danger signs, such as cracks, bulges and damp patches, and examine the ceiling carefully for damage from both above and below. Danger of collapse may not be immediately obvious, so, if you have any doubts at all, or the building has been subject to pipe bursts, melted-snow damage or vibrations from heavy traffic, it is worth getting it checked out by a professional. Problems may extend further than discoloured or damaged plaster: the laths and battens, the nails or even the joists may be faulty. Sometimes the plaster loses adhesion with its supporting laths.

Less serious on undecorated plain ceilings is crazing; providing the key is still sound, this can be disguised with lime putty, lining paper or a covering of fine muslin put up with flour paste and glue. Finish with a coat of white limewash or paint. Larger cracks can be resolved on an otherwise sound ceiling by widening to where the plaster is well attached, undercutting the edges, wetting the area and replastering with superfine plaster of Paris. A bulging ceiling can sometimes be pulled back into place and screwed to the joists (safer than nailing), using brass discs countersunk into the ceiling with double washers. Plaster of Paris can be used as a filler, but the repair will always be visible even if you skim coat and paint afterwards. Where a ceiling has collapsed and the laths are damaged, you will have to replace with wood or metal laths (page 57) and replaster. Avoid using modern plasterboard where possible. If a section of ornamental ceiling has

fallen but is undamaged, you can put it back up using wet plaster or screws with countersunk heads. Missing details can be remoulded and replaced by experts, although this is expensive (see above).

Although plasterboard plus skim coat has been the most common practice when plastering a ceiling since the nineteenth century, the traditional method for a period home involves three layers of plaster keyed to each other and the first scratch or lathing coat applied between the laths to form hooks. Wooden laths have been in use since the sixteenth century; the metal mesh was patented in 1841. If replacing wooden lathing, the timber should be pressure-treated and galvanized nails should be used. Traditional plaster comprises the usual ingredients of lime, sand and animal hair, sometimes with the addition of cement or plaster of Paris to speed up the setting time. Each successive coat is progressively weaker.

TOOLS FOR STONE AND PLASTERWORK

Although the tools for stone and plasterwork have changed little over the centuries, they are often made from modern materials, and this can change the effect of the final finish. Look out for original old tools at car-boot sales and in junk shops; there are also dealers specializing in old tools.

Beechwood mallet: a large wooden mallet that does a heavy-duty job, yet absorbs much of the impact of hitting the stone.

Bolster chisel: heavy-duty chisel used with a hammer for knocking out damaged sections of mortar, stone and plasterwork.

Claw boaster: type of heavy metal chisel used for dressing hard stone.

Club hammer: heavy hammer used with chisels to remove damaged stone and plasterwork.

Section Two

Cold chisel: used in conjunction with a club hammer, a cold chisel is useful for knocking out damaged sections of plaster or mortar.

Devilled float: wooden float punched with nails used for scoring and keying a plaster or render surface.

Drags: metal toothed tools for scoring flat surfaces, and available in different sizes and shapes.

Dummy: metal-headed (lead and zinc) mallet, used in conjunction with the wooden-handled mason's tools.

Filling knife: the flat flexible blade of the straight-edged filling knife is useful for all types of general repair work where filler has to be spread into holes in masonry or plaster. Blade widths are available from 25 mm/1 in to 100 mm/4 in.

Handsaw: useful for cutting off awkward pieces of stone when shaping.

Hawk: the hawk or spot board is rather like an artist's palette, although it tends to be square, and nowadays is made of aluminium. It holds the plaster or mortar while working. Sizes range from 250 mm/10 in square to 350 mm/14 in square.

Heavy pointed hammer: sometimes used to make decorative marks in the surface of the stone.

Ladders and scaffold: some kind of ladder or scaffolding will invariably be required to reach walls and ceilings safely, both inside and outside the building. A stepladder is adequate for low-ceilinged properties, and folds away easily for storage. Stepladders are available in wood or aluminium, but the aluminium types are lighter and easier to maintain. Both come in a range of sizes. Ladders can be single or extending to

reach the required height. You need at least a two-rung overlap for heights up to 4.3 m/14 ft and a three-rung overlap for reaching heights up to 5 m/16 ft. Again, ladders can be timber or aluminium, the aluminium being easier to handle and maintain. To put up a ladder, place it at the bottom of the wall to be scaled and, working your hands down the rungs, slowly but firmly push it upright. The correct angle should see the base resting at 70° to the ground, or a quarter of the height of the ladder from the wall. Many modern ladders incorporate an indication arrow to show the correct angle. Always follow the safety guidelines below when using a ladder.

- An aluminium 'stand-off' or ladder-stay holds the ladder away from the wall and improves stability.
- Beware of overhead cables.
- Ensure that the ground beneath the ladder is stable. Modern ladders often have rubber feet. If the ground is soft, use a board or proper base units to prevent the legs sinking in. A sack of sand behind the uprights is also useful for extra stability.
- There should be someone at the bottom to anchor the ladder if extended over 3 m/10 ft.

Using a ladder-stay makes safety sense.

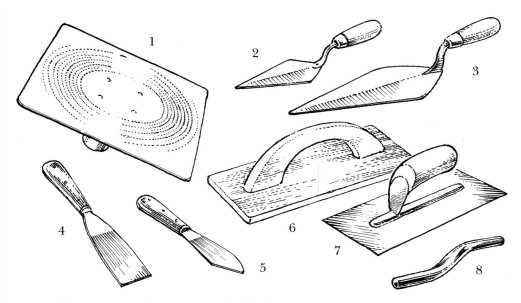

*Plastering and pointing tools: 1 plasterer's hawk; 2 pointing trowel;
3 brick trowel; 4 filling knife; 5 putty knife; 6 wooden float;
7 plasterer's trowel; 8 brick jointer.*

- Always carry a ladder safely: the lengths should be supported on your shoulder with most of the weight to the front.
- Do not rest the top of the ladder against a guttering: use a ladder-stay.
- Do not use ladders when the weather is very windy or in frosty conditions.
- Always wear shoes, not thick-soled boots or wellingtons, and avoid flapping trousers (or skirts!).
- If you do not like heights, do not look down or lean out too far.

Scaffold towers can be bought quite economically, or you can hire them by the day or weekend. Folding systems slot or bolt together, which makes storage easy, and are useful for both interior and exterior

work. Scaffolding is far easier and safer to use than ladders when working on high walls or ceilings or stairwells. Sectional or tower scaffold generally comes in sections up to 9.6 m/32 ft high with a platform size of 2.4x1.2 m/8x4 ft. The tubular sections slot together to make the framework as required. For a large area, two towers are built and scaffold boards balanced between them. Scaffold height should never be more than three times the dimension of the base. For reaching lower heights, a wooden trestle (like a low platform or table) may be adequate; this usually folds away for easy storage.

Mason's callipers: measuring tool for making exact measurements of uneven features.

Mason's saws: various handsaws are available for cutting soft stone. Some have a special blade the angle of which may be varied for making awkward cuts.

Pitching tool: kind of cold chisel with a blunt, broad edge, used with a mason's mallet to chip out a rough shape in stone.

Plasterer's trowel: now usually made of steel, the plasterer's trowel is used to apply and smooth over plaster on walls and ceilings. Sizes vary from 100x250 mm/4x10 in to 125x275 mm/5x11 in.

Pointing trowel: the traditional-style trowel for applying mortar is generally available in blade lengths of 160 mm/6½ in to 300 mm/12 in and may be bull-nosed or pointed, depending on the finish you require.

Spotboard: clean piece of board used for standing the mixed cement/mortar on.

Stiff brush: useful for removing dust and debris before undertaking any repairs to stonework or pointing.

SECTION TWO

Raising the spotboard on an old table or low platform makes it easier to use.

Tyrolean machine: hand-cranked to produce a fine spray of Portland cement/aggregate, giving a textured finish on walls.

Waster: type of chisel used for fine shaving of the stone.

Wooden float: produces a fine matt finish on the surface of skim plaster or cement mortar. It usually measures 125x275 mm/5x11 in. It is more weather-resistant and more in keeping with period properties than the finish produced by a modern steel float.

Wooden-handled chisel: the wooden handle absorbs any impact and offers better control, so this is ideal for intricate work on soft stone. Often used in conjunction with a dummy.

Wooden-handled gouge: used for intricate shaping work in soft stone.

OTHER USEFUL TOOLS

- Spirit level
- Cement mixer
- Buckets
- Wheelbarrow
- Sandpaper
- Set-square

STONE AND PLASTER PROJECTS

PROJECT ONE
LAYING FLAGSTONES

Original flagstone floors are frequently found still *in situ* in older country properties; the hallway, kitchen or pantry are the most likely locations, but you can find them laid throughout the ground floor, including sitting rooms. They might be marble, freestone (such as York stone), granite, slate or quarry tiles. Often they will be hidden under another, later flooring surface, and it is worth exposing them and preserving the flags wherever practical. It does not matter if they are uneven and worn – this gives character and a mellowness to the feature; often, however, slabs will be cracked or chipped, or may have lifted to provide something of a hazard. Levels can be adjusted by lifting and putting a weak mortar mix beneath. Restoring a flagstone floor is hard work, because the flags are heavy, but the project is a relatively simple one. Take care how you lift and carry the slabs, using a trolley to transport them more conveniently. You can replace damaged

stones quite easily, if not from other parts of the house then from reclamation yards, which stock a wide range of different types of stone and slate slabs to enable you to get a good match. They are usually sold by the square metre/square yard. The flags may be laid in a regular pattern or be different shapes and sizes laid in a random manner.

The most effective way to clean flagstones is to scrub with a hydrochloric acid solution; this is the best way to remove old cement from reclaimed slabs. The acid is highly corrosive, so be extremely cautious how you store and use it: always wear protective clothing and use in a well-ventilated room – outdoors for preference.

Repairing the floor may be a matter of simply relaying any loose slabs and replacing the odd one that may be damaged. This might be a good opportunity to rearrange some of the slabs in high-wear areas to the room perimeters to even out the wear. More extensive damage may necessitate the whole floor being lifted and relaid. This also might be the ideal chance to put in a damp-proof course, as slabs are often laid straight onto the soil in old unrestored properties; after installation, the bedding layer of sand can be laid directly on the plastic membrane. If you do need to dig out the old floor before relaying the flags, you may like to consider putting in a system of underfloor heating to replace radiators. This is an excellent way of heating a period property, as it is unobtrusive yet extremely efficient under a hard-surface floor. Heat is supplied via hot water pipes under the subfloor, and these can be run from any usual fuel source (gas, coal, wood, electricity) like any other system. The heating is completely controllable via a central automatic control and individual room thermostats. Installation is usually carried out by a specialist. If you are digging out and relaying the floor for any reason, remember to calculate the final height, or you may be effectively lowering your ceilings and making your doors impossible to open!

1 If simply repositioning the odd slab, lever up carefully with a crowbar and gently knock off any mortar around the edges of the flag.

Clean up any crumbled mortar and other debris from the cavity. Check the level of the sand and top up if necessary with sharp sand, before gently lowering back into position. Allow to settle and remortar the joins.

2 If you are relaying the whole floor, it may be a good idea to number the slabs; they can be stacked vertically against a wall – not too far from where they are to go, to save a lot of hard lifting and carrying. First check that the sand is level, using boards and a spirit level. Flagstones tend to be of uneven thickness, so the sand will have to be adjusted slightly as you lay each slab. Regularly sized and shaped flags can be laid in patterns, but irregular sizes must be laid randomly – the latter process involves a lot of trial and error until it looks right. Remember to stand back frequently to help judge the effect you are creating. Some flagstones fit snugly together and should be butted up together; a piece of timber and a mallet are useful for knocking the slabs more closely together. Alternatively, leave a space between for mortaring. Size of the joint will depend on the size and style of the flags, but do not make it too wide – about 13 mm/½ in is usual. The effect should be that of a stone floor, not predominantly cement.

Trial and error may be required before randomly laid flagstones look right.

An old flagstone floor is worth maintaining in a suitably period property.

3 Sometimes it will be necessary to cut a flag: if you can, lay all the complete slabs first, then cut all the remaining stone required in one go. You will need a stone disc-cutter, and these can be hired by the day or weekend. They make a tremendous amount of noise and dust, so you will have to do your cutting outside – check the weather forecast before planning this project! Safety equipment, such as a dust mask and safety goggles, is essential, and this is often supplied with the machine. If you have not used one before, it would be a good idea to

practise on a waste piece of stone, as the machine can easily slip or kick back. Mark the cutting line on the flagstone in chalk or pencil, after double-checking your measurements, as replacement stone is expensive. More complicated shapes may require making a paper or cardboard template first. After cutting, clean up the stone and fit as before.

Cutting a flagstone is a noisy, dusty job, best done outside.

Dry mortar may be brushed into the joints, then carefully watered in.

4 When all the flags are laid and you have double-checked with your spirit level that they are level, it is a good idea to leave them to settle down – about a week should be long enough. Then the joints can be mortared to finish off. Pack them with a stiff 4:1 sand/cement mortar with a pointing trowel, taking care not to smear any mortar on the surface of the stone; if it does get spread about, wipe away immediately. Alternatively protect the edges of the slabs with 75-mm/3-in masking tape before you start. If possible, avoid walking on the floor – or at least the joints – until the mortar has hardened.

5 A stone floor is naturally very durable and is easy to clean simply by sweeping and washing. Oily stains can be removed with a proprietary patio cleaner if necessary. It is not usually recommended that you seal an old stone floor with plastic-based sealants, as they prevent the surface from 'breathing', but you can polish up with a beeswax-based polish if you have the energy and inclination. Usually, years of feet travelling backwards and forwards produce a deep shine and pleasant patina.

PROJECT TWO

STAIRS AND STEPS

Stone steps and stairs do exist in early cottages – you sometimes see them located on the outside of the house. It was not until the late seventeenth century, however, that the staircase became an important architectural feature and focal point of the grand entrance hall. Stone steps often had ornate wrought-iron or timber balusters and, despite centuries' wear and tear, you usually find them in good condition. Where the stone has been chipped and cracked sufficiently badly to spoil the overall appearance, sections of stone can be matched and patched, or small chips repaired; there are various epoxy-based adhesives available which can be used to reassemble broken stone details or to resurface small areas of stonework that have worn or weathered. Stone or marble steps up to the front door are far more common, and, again, although damage is not frequently seen, they are repairable. A flight of stone steps also makes an attractive feature in the garden: in the formal period garden they might be wide and shallow, perhaps elegantly curved into a semi-circle and grandly flanked by carved balustrades or even statues perched on pedestals. This is the kind of external decorative staircase used to link garden to patio or terrace. Simpler stone steps have their place in the more informal, natural garden, too: stone treads might be irregular slabs or shaped

pavers teamed with maybe brick or timber to link one level to another, and partially hidden by surrounding foliage to soften the effect. Even a few such steps disappearing from view, or leading to a new, unexplored part of the garden, add a sense of interest and mystery and give a small garden the impression of being larger than it really is. Collapsed or tilting steps can be dangerous, so should be dismantled and rebuilt as soon as the damage is detected. Slippery steps can be equally hazardous, so do keep the steps well maintained by removing any build-up of mud and algae with a stiff brush at least once a year.

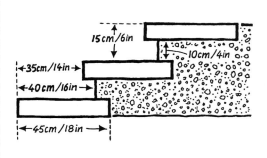

Planning the number and size of steps helps to calculate materials.

Stone treads with brick risers make an attractive combination for informal garden steps softened by plants.

1 If you are completely replacing a series of garden steps or building from scratch, you must first work out the proportions of the individual steps. You will need to know how many steps you will be building and what size they will be: the height of the risers and the depth of the treads. To make the calculation, you should estimate the total length and height of the proposed flight and divide this into an equal number

SECTION THREE

Above: Stone steps (and staircases) are often flanked by ornamental balustrades that may need cleaning, or minor repairs. Below: Even a few steps add interest and elegance when constructed in stone.

of steps. There are no set rules as to size. Generally speaking, an ideal riser height would be around 10–18 cm/4–7 in – something as high as 30 cm/1 ft would be uncomfortable for most people to negotiate, for example. The steps themselves might be around 45 cm/18 in deep with an overlap of 5 cm/2 in. Sometimes, where the site is irregular, it can be difficult to calculate. In this case, it is best just to start at the bottom with the first step, installing a 18-cm/7-in riser, and work upwards, making any necessary adjustments as you go. When you know roughly what size the steps will be, you can organize the materials: stone and slate make good treads, and risers might be constructed from brick, blocks, stone or paving fragments pressed into cement.

2 Working to your calculations where appropriate, dig out the first few steps roughly to size and see how they work in practice. You must always start at the bottom of a flight. Dig out the footings for the bottom step, 10 cm/4 in deep. Install the first riser and then the first tread, laying stone on a thin layer of sand to bed it in. A spirit level will help ensure that the steps are level, but a very slight incline is useful for encouraging rainwater to run off. Repeat with the next step and so on to the top of the flight.

PROJECT THREE
REPLASTERING A WALL

If the indications are that your plaster is not sound – a hollow sound when tapped means the key has given way – the wall will have to be completely replastered. Replastering is not as daunting for the novice home restorer as you might think; in fact, it is the process of knocking off the old plaster and preparing the wall surface that can be time-consuming and messy. Internal walls are usually plastered in two coats: the first, called the backing, browning or floating coat, is thicker and is intended to even out the wall surface. The second, finishing coat is

thinner and finer to produce a smooth finish suitable for paint, paper, panelling or tiles. Today, you can buy the plaster ready-mixed in powder form – all you have to do is add water. Cement-based plaster takes longer to set and does not key as well, but it is inexpensive; gypsum-based plaster is more expensive but quicker-setting, and is usually preferred. If you are a real traditionalist and want to do it the hard way, you can mix your own plaster (page 39).

How much plaster you need will depend on how thick you will be making the coats, and this can be a little unpredictable for a first-time plasterer. However, as a rough guide, an area approximately 90m²/110 sq. yd will require about 60 kg/132 lb of base plaster mix and 20 kg/44 lb of finishing plaster. Ready mix is sold in bags of 10 kg/22 lb and 50 kg/110 lb sizes. If you have never applied plaster to a vertical surface before, it might be a good idea to practise the technique before you tackle a whole wall. Simply mix up a bucketful of plaster and play around with the hawk and trowel (see below) until you feel confident when handling the plaster and the tools. You will also need to practise getting the consistency right: too sloppy and you will find the whole lot slumps to the floor just when you thought you had made a good job of it. The plaster can easily be knocked off the wall before it dries so that you can start the job again properly.

1 Begin by preparing the wall: all the old plaster will have to be hacked off using a bolster chisel and club hammer, or whatever comes to hand. If it is in bad condition anyway, this should not be too difficult, but it will make a tremendous mess; so empty the room of furniture and cover up furnishings wherever possible. You can hang dust sheets at the doors and change your footwear going in and out of the room, but, even so, you will find the dust manages to get out and settle all over the house.

2 Rake out all the loose plaster and clear away any other debris, then dust off with a stiff brush. By now, you will be able to assess how much

damage there is to the actual wall surface and to the laths. Broken laths will have to be repaired; or you can replace with more modern expanded metal mesh if you prefer (page 39). An old brick surface beneath may need repointing before you can start plastering. If there is a patch of smooth surface such as concrete or timber, it will have to be scored or scratched in some way to provide a good keyed surface for the plaster to adhere to. A cold chisel should do an adequate job of scoring criss-cross lines in concrete; a timber lintel is best covered in a section of expanded metal mesh, which is easily nailed on.

Nailing on expanded metal mesh provides good key for the plaster.

3 Gypsum-based plasters come in different grades to suit different types of wall surface which have variable absorbency. Some surfaces, such as brick, are highly absorbent and will suck the moisture away from the plaster, so will need spraying or soaking with water before work starts. You can use an old paintbrush or plant sprayer. The dryness of the weather and low humidity due to central heating will also make a difference.

*Where internal plaster is damaged, or has lost key, it
must be knocked back to the wall.*

4 Once the wall is prepared and you have assembled all the necessary
tools and materials, it is time to start plastering. It is important to allow
yourself time to finish the job, so never try to rush it. Unless the room
is a very small one and you are using a slow-setting plaster mix, it is a
good idea to divide the area to be plastered into sections or bays using
vertically fixed 10-mm/$\frac{2}{5}$-in battens about 1 m/3 ft apart. If you knock
the fixing pins into mortar, not brick or stone, they should pull out
easily. The battens should be fitted the full length of the wall, and

make a useful additional guide for levelling off the plaster, so check they are straight, packing behind them if necessary where the wall is very uneven.

5 Mixing the plaster correctly is essential to success. For the beginner, it is probably best to mix it up a bucketful at a time; the professional will use a wheelbarrow or trough, but then he/she will be using it up more quickly too. It is important that your bucket is clean and that the water for mixing is fresh. You are aiming at something akin to the consistency of porridge, with no lumps. Half-fill your bucket with water, then sprinkle on an equivalent volume of powdered plaster, running it through your fingers to make sure there are no lumps. When the plaster has filtered down into the water, stir it slowly with a clean stick until it is smooth and quite stiff. Err on the side of caution when adding the plaster: a sloppy mixture can be improved with a little more plaster, but water cannot loosen an over-stiff mix. Tip out the mixture onto a clean piece of board (called a spotboard) and chop and knead the plaster with a plasterer's trowel, until it seems workable.

6 When the plaster is ready, transfer a dollop from your spotboard onto a hawk, which you hold in the other hand, rather like an artist's palette. Push the plaster around on the hawk with the trowel until it forms a neat mound. Now place the trowel on the hawk with its blade at right angles to it and, tilting the hawk gently away from you to encourage the plaster towards the trowel, start bringing the trowel up towards the mound of plaster. As the hawk reaches a vertical angle, neatly push up the trowel and lift the plaster onto it. Keeping the trowel horizontal, you should be able to return the hawk to a horizontal position and tip the plaster back into the centre of the hawk. This semi-tossing action helps 'work' the plaster to a good consistency, just as kneading elasticates bread dough.

The next steps are described for a right-handed person; left-handers will begin in the left-hand corner and work left to right.

The first, floating coat is applied with a firm, sweeping movement.

7 Begin the first coat in the bottom right-hand corner of the first bay. Scoop about half the mound onto the trowel. Rest the edge of the trowel on the batten and tilt the blade upwards until the face is approximately 30° to the wall surface. Now push the trowel firmly upwards, tilting the blade more steeply as you press the plaster onto the wall. Try to use the trowel in a sweeping movement as you press. Scoop up the rest of the plaster off the hawk and repeat the procedure.

8 Keep applying plaster in this way, working across to the left-hand batten. Apply a second band above the first, working in the same direction, and repeat until you reach the top of the wall. Use a trestle to reach the top of the wall easily. Smooth off with the trowel and, as the plaster begins to set, place a wooden rule across it and draw it slowly upwards with a side-to-side action to level it off. Any excess plaster can be put back in the bucket to be used to fill in any hollows; then rule off again. When almost hardened, key the surface for the next coat using a nail or a devilled (nail-studded) float.

9 Now move onto the next bay, plastering it in exactly the same way. Leave the battens in position until you have finished, then go back and remove them and fill in all the gaps. If you do not have sufficient battens to divide off the whole room, remove the right-hand batten as each bay is finished and use the raw edge of the plaster as your guide.

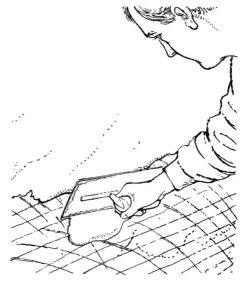

Key the plaster with a nail or the edge of your trowel before applying the next coat.

The finishing coat of plaster is smoothed to an even finish.

10 The finishing coat can be applied as soon as the floating coat has hardened; this will depend on the weather and type of plaster used. This time, mix the plaster to a relatively runny consistency similar to that of thick yoghurt. Keep this coat thin and even, beginning in the bottom left-hand corner of the wall and working with a bold, vertical sweeping movement. When you have covered a band around 2 m/6 ft wide, then go back and apply another, even thinner coat over the first.

Any ridges or splashes can be smoothed away with a wet trowel: sprinkle the blade with water and, holding the trowel at about 30° to the surface, stroke it lightly downwards. Move on to complete another band until you reach the ceiling. Overlap each band as you work to avoid ridges, although these can be smoothed away with your wet trowel. Finish by smoothing off with the trowel.

The finished surface should be smooth enough for paint, paper or tiles.

STONE AND PLASTER PROJECTS

EXPOSING AND RESTORING AN OLD FIREPLACE

Now that the fashion for ripping out and bricking up old fireplaces has passed, many homeowners are reinstalling or exposing them again, not just as a valuable architectural feature and focal point for the room but also as a source of homely warmth and welcome. Few are going to reinstate a fireplace purely as a decorative feature, although in latter centuries they did become an integral and important part of the overall room design (page 11). The chances are you will be equally keen to see a live flame in the home once more, if not an open fire (not renowned for its convenience and heat efficiency!) then an attractive wood-burner or maybe one of those real-fuel-effect gas fires which you can turn on at the flick of a switch. This is why it is important, before you start any ripping out or expensive restoration

This old fireplace had been bricked up. Knock out the bricks carefully to avoid damaging the structure beneath.

work, that you check out the safety and efficiency of the chimneys and flues. Even if they have not been taken away or blocked, they may be worn and potentially dangerous. It is worth consulting a chimney specialist if you have any doubts; they can carry out a smoke test. This will reveal if poisonous fumes are leaking through brick or stonework into the rest of the house. Chimneys may need rebuilding or lining, which will add to your restoration costs.

With the twentieth-century obsession for draught proofing, many rooms no longer have sufficient draught for a fire to burn efficiently. Surrounding trees and buildings may also be higher than they used to be, affecting the efficiency of the chimney. These problems can be rectified by altering the fireplace, the chimney, or installing an electric chimney fan, but they need considering in the early stages.

Opening up a concealed fireplace has that element of surprise and excitement not found in many other home restoration projects. You do not know what horrors or treasures you may find: the fireplace surround may be completely missing, or the fireplace may be small and rather insignificant. Then again, the installation may be much older than you think, revealing a magnificent carved surround, or an ancient and huge inglenook that will significantly increase the size and character of your room. Sometimes, there are several fireplaces, one behind the other, spanning different periods and centuries. In this case, you will have to decide whether to sacrifice a fireplace of a later period in order to expose a much older one. Whatever your decision, you must tread carefully to avoid damaging any valuable features.

1 Once you have established the presence of a hidden fireplace, it should be carefully exposed. Sometimes this involves little more than tearing down boards, but many have been bricked up. Tackle the bricks with a chisel and hammer, taking great care not to damage whatever is beneath. You should aim to preserve wherever practical any additional period features, such as a bread oven, shelves or recesses intended for storing salt, spices and tobacco, or a curing chamber.

2 When the fireplace is fully exposed and the debris has been cleared away, use a stiff brush to clean as much dust and dirt as possible from the exposed brick or stonework. Then assess the damage and decide the extent of the restoration work involved. It is likely that some replastering will be necessary; where you do not think there will be sufficient key, expanded metal mesh can be nailed on. Always use galvanized nails (which will not rust) and knock them into the mortar rather than into stone or brick. Choose nails of sufficient length to hold the mesh – up to 100 mm/4 in if necessary.

Applying the first coat of plaster to the arch.

3 Make up a strong bonding plaster mix following the manufacturer's instructions; bonding plaster is fibreglass-based and has particularly good adhesion. It is also relatively quick-drying, so do not mix up more than you can handle, especially if you are still hesitant about your plastering technique. When the plaster is well mixed and of the right consistency, follow the procedure for internal plastering as described in Project Three: load your hawk from the spotboard and work it over with the trowel. With around half the mound on your trowel, apply it to the wall with an upward, sweeping motion, pressing firmly. Start at

the bottom right-hand corner and work upwards and then across, smoothing the plaster with your trowel as you go. Drawing a wooden rule up the plaster as it starts to harden will help level it and remove any excess (page 60). Do not worry about getting the upright edges absolutely straight, or the arch perfect: this would look odd and unnatural in a period feature which should have some softness and irregularities in imitation of natural wear. Before the plaster sets hard, score it with the edge of a trowel.

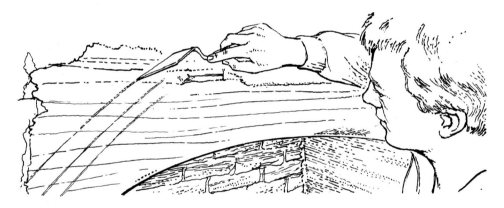

Scoring the plaster with the edge of the trowel to key the surface for the next coat.

4 When the bonding coat has hardened, apply a finishing skim coat of plaster in the same way, although using a thinner mix as recommended (page 61). Ensure that this final coat butts up neatly with the existing wall finish either side of the fireplace. Smooth with a wet trowel to achieve a level, seamless finish.

5 Here it has been decided to leave the back of the fireplace exposed as a feature; in a simpler, cottage-style property, treating the brick or stonework and lintels of open hearth fireplaces in this way can have a suitably rustic appeal. Repair and repoint any damage to feature brick

*The finished fireplace not only restores an attractive
focal point to the room, but could be fully functional too.*

or stonework, taking care to match the colour of the original as closely as possible. This means careful choice of sand and possibly the addition of a coloured tint (see 'Mortar', page 28). The pointing should be kept indented.

6 You will also probably have to refloat a new concrete hearth before installing a new grate or heavy cast-metal wood-burner. A wooden batten positioned along the front of the hearth will keep the concrete

Floating a new concrete hearth.

in place and create a reasonably level edge. Rub with soap so that it removes easily after the concrete has set. Mix up sufficient sand: cement mix and apply, about 38 mm/1½ in thick, with hawk and trowel. Smooth off with a steel float and leave to harden for around a week, depending on the weather, before installing any kind of grate or wood-burner.

7 The fireplace surround should be finished in a style to suit its size, position and period. Inglenooks and small fireplaces usually have a fairly simple finish, and the new plaster will only need painting to match the rest of the room. More ornate styles since the seventeenth century have featured glazed tiles up the sides in a slip panel, or sometimes over the whole fireplace area, inside and out. You can still buy the tiles; if you cannot find a complete set from one of the reclamation yards, then the original designs are still being made, from pretty blue-and-white Delft to Victorian heavy-relief styles, richly patterned florals or delicate Art Deco. A vertical set might comprise a flower, stem and leaves run over several tiles. Alternatively, you might

be seeking to replace a fireplace surround, and again these are available both as original features and faithful reproductions ('Style Guide', page 10). The wooden surrounds are usually drilled and screwed to the wall: a plug of wood is usually supplied to fill the hole for a near-invisible finish. Cast-iron models usually incorporate screw-holes for fixing; marble and stone surrounds employ a system of wires set in the back of the stone. These are screwed into the wall, then plastered to fill in the gap between wall and surround.

PROJECT FIVE

RENDERING AN EXTERNAL WALL

Depending on the local climate and how exposed the wall is, rendering can be weathered or damaged to the point where it ceases to do its job properly and lets in the wet. Once this has happened, you have no choice but to repair even a smallish section (see 'Repairing Render', page 88) or you will be laying the foundations for bigger problems in the future. Look out for surface crazing or delamination due to efflo-rescence of salts and frost; if the render sounds hollow when you tap it, you know you have lost key and the whole lot will have to come off. Modern cement-based render tends to set too hard for period build-ings and a traditional lime render recipe (page 35) is preferable for older houses. If the house is listed as being of historical importance, it may even be obligatory, so do check with the experts first. Generally, the render comprises three coats for walls in a relatively exposed posi-tion, each coat progressively weaker than the last, and with care taken to key each surface between coats. Never tackle rendering in frosty weather or when the atmosphere is hot and dry, as the render does not set properly. Good preparation of the wall is essential for effective adhesion. Also check that access is safe and adequate before you start the job; scaffolding is the safest, most convenient way to reach the upper storey (pages 44–5). If the render is not applied correctly or in

SECTION THREE

A plain rendered finish is usually painted for added protection and aesthetic reasons. Natural pigments were once used; today, modern masonry paints are available in a wide range of colours.

one smooth operation, it will not do its job properly; so do practise first, on the garden shed or an old bin, if you have never done it before.

1 First remove all the old render, or what remains of it if the damage is extensive. A cold chisel or bolster and a club hammer should make short work of any stubborn sections. Do not forget to work out how you are going to dispose of the debris once it is down; it would make useful hardcore for the foundations of a patio, for example. Once the wall is exposed, it should be thoroughly cleaned of dirt and dust with a stiff brush – this is essential if you are to get good adhesion of the render. Check the condition of the wall beneath. Any large holes in stone and brick will have to be filled to improve weather resistance (see page 36). A waterproof additive in the pointing mortar would be a good idea for

a long-term protection if the site is an exposed one. This is usually added to the water before mixing. Leave any smaller holes and cracks, as they will be filled by the initial scratch coat, and are useful for providing extra key. The area to be rendered is usually sectioned off with vertical battens, just as you would for plastering an internal room; this not only divides a large area into more manageable portions, but also helps keep the render coat level (page 58).

2 Just before applying the render, the wall should be wetted with water; this is especially important during dry, warm weather, as it is intended to prevent the wall absorbing moisture from the render and affecting its setting abilities. Use an old paintbrush or even a hosepipe spray for large areas.

3 Start mixing your first render coat, making sure that you measure the ingredients accurately and that all your tools are scrupulously clean. Your choice of ingredients will depend on the age and type of property and how historically authentic you want to be. Not everyone fancies mixing cow dung or cheese into lime render to improve its workability! A more modern cement-based render tends to average around 4:1 sand:cement, although it will be weaker or stronger depending on which coat you are applying (see page 73). Whatever your choice, it is absolutely vital that it is mixed thoroughly into a thick, smooth, creamy consistency. Unless you are tackling only a small repair, you are advised to borrow or hire a cement-mixer for the job; if this is electrically run, you will have to organize a convenient supply too, probably via an exterior-grade extension lead. Be careful when operating electrical machinery while handling water.

MIXING BY MACHINE

Start with half the sand and some of the cement or cement-and-lime. Add water and cement or cement-and-lime in turns, finishing with just

enough water to make it workable. It should not be sloppy, nor collecting in clumps as the machine turns.

MIXING BY HAND

First mix the dry materials thoroughly; form a crater in the mound and add some of the water. Start working the dry materials into the centre then turn it over and add more water slowly, mixing thoroughly as you go in order to reach the correct consistency. A watering can with a fine rose is useful for controlling the final amount of water added.

4 Once the wall has been wetted (but not soaked), begin rendering by taking a load of render mix from the mixer to the wall in a wheelbarrow. Transfer a mound to your spotboard and, from here, load your hawk as described on page 59. Work the render on the hawk with your trowel, using the same technique as for internal plastering; then, starting at the bottom right-hand corner, take about half the hawkful of render on your trowel and press it onto the wall in a sweeping upwards motion. Use the trowel to smooth and level, working a section of around 1 m/3 ft at a time. Aim at a coating around 8–12

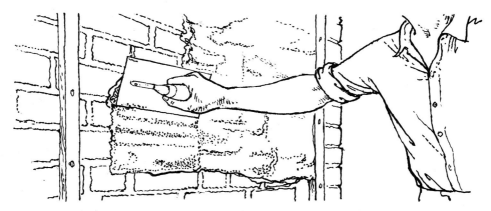

The render is applied to the first bay with wide, firm sweeps of the trowel.

mm/^{3}⁄$_{10}$–½ in thick – never more than 15 mm/^{3}⁄$_5$ in anyway. Comb or scour the render before it hardens, in preparation for the next coat.

5 When the scratch coat has hardened, you are ready for the next layer: the floating or backing coat. It should be mixed slightly thinner than the first coat, and, if you are using a cement-based render, be less rich in cement so that the mix is weaker. Again, it is essential that the render is mixed thoroughly and kept free from any dirt or debris. Apply it to the wall in exactly the same way, smoothing over and scoring to key the surface when the job is finished. Leave to harden for about a week.

The surface must be scratched or keyed to help adhesion of the next coat.

Once the render has set, the battens should be removed and the gaps filled with fresh mortar.

6 The final coat is called the finishing or skim coat. Mix and apply to the wall as before, again making the mix slightly thinner and weaker than the second coat. It should be around 5–8 mm/⅕–^{3}⁄$_{10}$ in thick. Finish with a traditional wood float to produce an appropriately flat, slightly textured finish (a modern steel float does not create the same

effect). Older buildings should have a soft, slightly undulating finish. The hardened render is often limewashed or painted (see below), which is important for providing extra protection, of course; but there are also many different-textured and decorative effects which can be applied to the render while still soft, and which have gone in and out of fashion over the years. Some are particular to a certain region and are rarely seen elsewhere; others are completely individual to a quirky homeowner. If you wish to attempt any of these effects yourself, it is a good idea to experiment first on a board offcut, until you are happy with the technique. This is particularly important if you are trying to match an existing effect.

7 Often the render would be marked or patterned with a variety of improvised and specialist tools while still soft: scoring an 'ashlar stone' finish would imitate the effect of stone blocks; a trowel blade and a straightedge might be used to score the area in a criss-cross effect; random sweeps of a wooden float produce what is called 'cottage texture'; toothed combs, blades and brushes can all be employed to make their mark. Even sackcloth might be dragged across the render to add texture and interest.

8 Roughcast – sometimes called 'wet dash' – involves actually mixing small stones in with the final render mix and throwing it onto the wall with a laying-on trowel. The usual mix is 1½ parts cement:½ part lime:2 parts shingle:3 parts sand. The resulting grey finish is usually painted.

9 Pebbledash or 'dry dash' tosses selected coarse aggregate such as spar or pea shingle onto the top coat of render while still soft. The stone varies in size from 6 to 13 mm/¼–½ in and is available in a choice of colours, which will affect the final finish as the shingle remains exposed and is not usually painted.

10 A tyrolean finish is relatively modern, and is applied using a special

hand-cranked machine which produces a fine spray of textured Portland cement:aggregate mix. Three coats are necessary: the first is put on diagonally downwards at an angle of 45° and working from right to left; the second is worked from left to right; the third and final coat is worked vertically upwards. it is important to keep the machine moving and the handle turning while in use, to prevent any build-up of materials.

Pressing shells into the wet render to create patterns and pictures is a painstaking process, but produces stunning results.

SECTION THREE

11 You sometimes see other material pressed into the render to create wholly individual and unusual homes: large coloured pebbles, shells and even small pieces of broken glass or crockery might be used to produce idiosyncratic decoration, making pictures and patterns in panels or completely covering external walls. The work is painstaking – just collecting sufficient material might take years – and, like pebbledash, can only be worked in small sections, applying a patch of render at a time.

12 Plain render used to be finished with limewash and tallow, and was an important part of protecting the external plaster from weathering. It was often coloured with umber or ochre; these earth-based pigments were found to be more resistant to the bleaching effect of the lime. You can reproduce the effect today, although there is also a wide range of modern masonry paints and finishes available. These offer both excellent weather resistance and a wide range of colours. However, as they are often cement-based, they can be too hard for traditional lime render, and it may crack and come away. If modern paint has already been used and cannot be removed, you will have to continue using it, as limewash will not adhere to it. Be careful what colour you choose, as some are totally unsuited to a period property. Others look out of place in the wrong location – seaside pink, for example, or the darker townhouse colours. Cement and masonry stone-based paints must be applied to a clean, dry surface to be effective and can be used with a special brush, roller or spray. The paints are abrasive so will wear out these tools quickly. For a heavily textured surface, use a 15-cm/6-in stiff brush for good coverage; on smooth surfaces you can use a roller. Load a 20-cm/8-in plastic or nylon roller from a paint tray and work evenly across the wall from left to right and then right to left, overlapping the first pass slightly. Aim to work in an even herringbone pattern. Finish off by working the roller vertically downwards, taking care not to let the brush drag on the surface and make unwelcome brush marks.

PROJECT SIX

FIXING A PLASTER CEILING COVE

Plaster coving is often fixed to the angle between the wall and ceiling in period properties; it makes a room look less square and more decorative, avoiding an ugly join. It is also useful for hiding the ugly (but not structurally damaging) cracks you often find at this junction, due to seasonal movement of the building. The coving might be elegantly simple or highly decorative, depending on the age, type and style of the property. Replacing damaged original period mouldings can be costly, as it all has to be hand-made to fit by an expert, especially if the feature is of particular historical interest. However, ready-made reproduction plaster coving has its place and is available in a choice of styles in 1-m/3-ft lengths. Coving is also available in cotton-fibre and polystyrene, but these are not really suited to a period-style restoration.

1 Begin by preparing the area to be covered: it should be clean, dry and free from any flaking paint or wallpaper. You can remove a strip of wallpaper by scoring with a sharp knife and peeling or scraping off – do not be tempted to wet the paper, as you will rip and damage more than you intend to. To mark an accurate horizontal guideline for the coving, snap a chalked stringline along the wall where the base will come to. Holding a piece of coving along this line will enable you to mark in the ceiling line, too. Tap nails above and below to provide temporary support.

2 Measure around the room and cut lengths to fit using a fine-toothed saw. Sometimes, fancy corner-sections are available; alternatively, on plainer styles, the coving has to be mitred to make a neat join at internal and external corners. Where a paper template has been supplied with the coving, this (or a mitre box) will indicate where the corner mitre should be marked for an internal or external corner. Cut along the line with a saw or sharp knife and smooth any rough edges

Above: Classical and other ornamental-style friezes are available by the metre and are easily installed

Right: A deeply moulded coving hides the join between ceiling and wall. It is traditionally painted white or a pastel shade.

STONE AND PLASTER PROJECTS

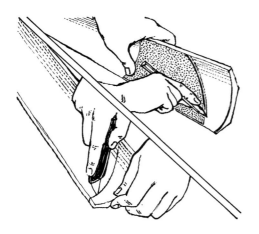

Corner shapes can be traced from a template and cut with a knife or saw.

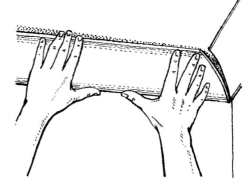

After applying adhesive, the section of coving is slid into place.

with sandpaper. The point at which your chalk lines on the ceiling intersect provides a guide to position when mitring the coving.

3 Once all the sections are prepared, the coving is installed by butting up the lengths along the chalked guideline; they are held in place using a special adhesive. This should be mixed to a smooth creamy consistency and spread thickly onto the back of the coving. If the adhesive does not grip firmly enough to hold the coving up without slipping, knock in a few temporary support nails and remove them when the adhesive is dry. If any excess adhesive oozes out from under the coving, re-employ it straight away to fill any gaps at the edges or along the join between lengths. The remainder should be wiped away with a wet brush before it dries.

4 When the adhesive has set (this usually takes around 24 hours), the coving can be painted, either to match the wall and ceiling finish or in white to contrast. Other colours are sometimes traditional (see 'Style Guide', page 14).

*T*IPS AND
*S*HORT *C*UTS

*W*hen it comes to restoring and renovating old properties, any attempt at faking it or taking a short cut would cause the purists to throw up their hands in horror. Although it is true that thorough preparation, the correct materials and adherence to traditional techniques are essential, not just for authenticity but also for safety, longevity and preservation of what remains historically, funds and opportunity do not always make it possible for a job to be done immediately. Provided that the basic structure of the building is made safe, and that any decay or deterioration is halted and you are not causing any permanent damage to an original feature, there is no reason why you should not employ a crafty cover-up or an authentic-looking alternative.

INTERNAL WALLS

Provided that a poor-quality internal wall or ceiling is not in danger of collapsing, and the condition is not likely to deteriorate if not treated promptly, there is no reason why you should not do the same as homeowners have done for centuries – cover it up until the problem can be rectified. Tapestry, timber panelling and to some extent

wallpaper have long been decorative ways to hide the walls, keep in the heat and reduce condensation in old houses. You can do the same where walls are chipped or stained. Tapestries are expensive, but rugs and dhurries make good wallhangings and add warmth and texture to a period room. An alternative is to use lengths of fabric on the walls; it makes an excellent and elegant cover-up, yet is easily removed. The material can be simply shirred onto narrow canes or wires, or stretched taut on special metal tracks. The same treatment is also a superb way to hide an ugly ceiling, especially when gathered from a central boss or motif to create a tented effect – a very effective treatment for cosy bedrooms and dining rooms. Wooden panelling is a more permanent treatment and an expensive one, even if you use reclaimed timber or even old panelling rescued from another property. Half-panelling to dado level is less imposing (and less expensive) and may be a good way to protect an old wall where it receives most damage. New tongue-and-groove panelling can also be fastened to a system of battens using pins or clips, then painted or stained to whatever period finish you choose.

Plaster Repairs

Provided that a wall is basically sound, smaller areas of damage can be successfully patched, although you cannot always guarantee that the repair will be completely invisible. The best cover-up once the repair is completed is wallpaper – or, where this is not practical or suitable for a simpler style period property, an old-fashioned paint finish, such as wall washing, sponging, dragging, stippling or rag-rolling, will make an excellent and inexpensive disguise for a less than perfect surface.

Warning

When considering any kind of finish for both external and internal period walls – paint, varnish or sealant – it is important that the finish still allows the wall to breathe and that it does not seal in a surface that you may later wish to remove, such as distemper.

TIPS AND SHORT CUTS

FILLING CRACKS AND SMALL HOLES

Small cracks and holes in plaster walls are easily filled and filed to a fine finish using an interior-grade cellulose filler, available in powder or ready-mixed form. You should first explore the extent of the damage and confirm that it is not a much bigger job than it first seems. Small holes are simply filled and smoothed off. Cracks should be scored with the side of your filling knife to widen the cavity, and any dust and debris brushed away. The crack will need damping with a wet paintbrush; then the filler can be packed tightly into the crevice with the knife, by drawing it firmly across the filler at right angles to the wall. If the crack is a deep one, it may be necessary to build up the filler in layers, allowing it to dry between applications. When filled, the filler should be smoothed off, slightly proud of the wall surface, and left to dry and harden for several hours. A rub-down with sandpaper will produce a smooth, flush finish. A badly crazed but otherwise sound ceiling can be disguised with lining paper, wallpaper or a textured paint finish (although the latter is difficult to remove). However, this treatment is not suitable for ceilings of any historic interest, particularly if there is evidence of decorative or applied plasterwork.

FILLING LARGER HOLES AND DAMAGED SECTIONS

It is possible to patch a wall without completely replastering, provided that you are confident the rest of the wall is sound. You should always gauge the extent of the damage: if a solid wall sounds hollow when you tap it, the whole lot will have to come off. To repair a hole in a solid wall, score round the edges with a nail or similar sharp implement to remove any loose material, and brush away the dust and debris. After moistening the hole with a damp paintbrush, push filler or plaster into the hole and allow to harden (a good stiff mix is best). Apply a second layer in the same way and, when this has set, sand it flush with the wall,

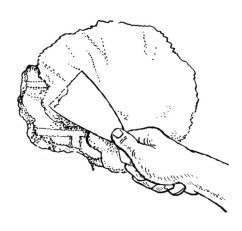

Repair to a solid wall.

Repair to a lath-and-plaster wall.

or use a broad-bladed scraper to remove what you do not need.

To repair a hole in lath-and-plaster, first cut away the loose plaster from the laths and brush away any dust. If the laths are damaged, they can be bridged using small sections of expanded metal mesh wedged into the base of the hole. Fill this with plaster to just below the surface, then cross-hatch with a knife-edge and leave to harden for about 30 minutes. Mix up more plaster, dampen the patch with your brush, and finish the filling. When this has dried, you can apply a thin top coat of plaster to be sanded flush with the wall when hardened.

PATCHES IN PLASTERBOARD

In a more recent property, or where a partition wall has been erected, it may be plasterboard rather than a solid wall that needs repairing. Since a cavity lies behind, the board should be patched rather than filled. Cut a piece of plasterboard slightly larger than the hole all round. Make a hole in the centre of the patch, and thread through a short piece of string with a nail tied to one end. If you smear a little

TIPS AND SHORT CUTS

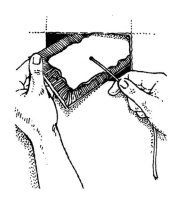

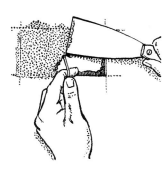

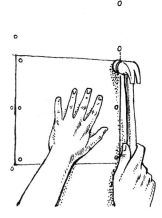

Repairing a small hole in plasterboard: inserting a patch behind.

Plastering over the patch while holding it in place with the string.

Applying an external patch to plasterboard.

freshly mixed plaster onto the side opposite the nail, you can use the string to guide the patch into the hole, plaster-side facing outwards. When you have manoeuvred the piece behind the hole, pull the string taut to straighten the patch and press more plaster into the hole to about 3 mm/$\frac{1}{10}$ in beneath the surface. Leave to harden, then cut the string, dampen the patch and add sufficient plaster to bring the hole flush with the main surface.

Larger holes, that is larger than 75 mm/3 in across, must be patched from the outside. The damage should be cut back in a rectangular shape to reach the supporting studs on either side. Cut a new piece of board to fit, then nail it to the studs using special plasterboard nails. To hold in the top and bottom edges, press finishing plaster with a trowel into the gaps or tape scrim (jute sacking) around the joins before smoothing over the patched area with a plaster skim coat.

REPAIRING A CHIPPED OR DAMAGED CORNER

The corners of walls are the most prone to damage, and are often chipped and ragged. A little chipping is easily remedied by several thin layers of filler built up until proud of the wall, then sanded down when the final coat has hardened. More extensive damage can be repaired using a batten to achieve a straight edge: although with older properties you do not want this to look too sharp and precise. Fix a wooden batten to one side of the corner, its edge flush with the other side. With a trowel and filling knife, one side can be built up so that it lies flush with the batten. When this has hardened, remove the batten and repeat the procedure on the other side. A final sanding will smooth and level the corner; round the edges slightly to soften, if necessary.

Repairing a corner using a batten.

EXTERNAL WALLS

Temporarily disguising a less than perfect-looking external wall is not quite as easy, especially if you do not want to alter or spoil an existing surface. Again, it is essential to determine whether the walls are sound or require more immediate and extensive repairs. Masonry paints and rendered finishes are available, and can instantly transform a property;

but you should also consider whether they are in keeping with its age and style, and also, should you wish to carry out repair work in the future, whether they can be easily removed. A temporary trellis, or wires fastened to the wall, planted with a fast-growing climber, is probably the outdoor equivalent to tapestry or wallpaper; check first that your chosen plants are not likely to damage wall or mortar. Where damage is only localized and the rest of the wall is sound, patching or part repair can halt the damage and improve the appearance, although the new work will often be discernible. Choice of the right colour and type of materials is vital, even for a small area. Some renovators like to smear new sections with yoghurt or dung and chopped mosses to encourage quick moss or algae growth and a rapid ageing effect.

REPOINTING A SMALL AREA OF WALL

Where mortar has become cracked and loose and is letting in rainwater, it is advisable to repair the section to prevent further damage to the wall structure. This work should not be undertaken in frosty or extremely hot, dry weather, because the mortar will crumble. It is important to keep the mortar mixture relatively dry, as, if runny, it will be too difficult to handle, and may run out. The usual mix for repairs is one part cement to one part lime to six parts soft sand; colour of the sand will determine the finish. All the old mortar must first be chipped out using a slim cold chisel and a club hammer, taking care not to damage the surface of the stones. When this has been done, rake out all the loose particles and brush the joints to remove any remaining dust and debris. Immediately before repointing the area, brush the joints with water using a household paintbrush to keep the mortar moist). Using a hawk and pointing trowel, cut off rounded pieces of the stiff mortar and press them into the vertical joints. Trim off the excess with the edge of the trowel, avoiding, as far as possible, 'buttering' the surface of the stones, and matching the finish of the

original pointing (page 28). When all the vertical joints are completed in this way, tackle the horizontal ones and leave to harden.

REPAIRING RENDER

Sections of render sometimes become loose and can be patched, although it may be difficult to make a completely invisible repair even if you are repainting the wall afterwards. All the loose parts should be knocked off, using a bolster chisel and club hammer. Mix up as much render as you think you will need (page 56) to a consistency rather like that of softened butter, making sure there are no lumps. Render sets quickly, so only make up as much as you can use in around 20 minutes. The wall will have to be well dampened with water immediately before work starts. Apply the first render coat with firm upward movements of the trowel to within 6 mm/¼ in of the remaining render surface. Starting at the bottom of the patch, press the lower edge of the trowel firmly against the wall, then sweep it smoothly upwards. Before the first coat sets, score it with the point of your trowel to key it for the next coat. The second coat can be applied after 24 hours when the render has hardened. This layer is worked from left to right and from the top of the repair downwards, towards the bottom. Draw a piece of straightedge timber, with its edges resting on the existing render on either side, upwards across the new render to achieve a level surface flush with the old. Finish off the patch with a wooden float.

REPAIRING PEBBLEDASH

Missing patches of pebbledash can be repaired by preparing the surface and applying two coats of render as described above. While the second coat is still wet, throw on small stones or shingle using a small trowel. If coverage is not dense enough, use a piece of board to press more stones into the render. Leave to dry. Painting the whole wall will help disguise the repair.

$\mathcal{G}$LOSSARY

Adze: hand tool used for dressing timber.

Aggregate: naturally occurring sand, gravel or crushed stone.

Architrave: moulding around a door or window, or the lowest part of an entablature that rests on the columns.

Ashlar stone: regularly shaped blocks of stone for fine building work.

Baluster: post or column.

Balustrade: ornamental rail supported by balusters.

Battens: sawn strips of wood used to provide support, restraint or a guideline.

Boaster: sharpened chisel used in masonry work.

Bolster chisel: heavy-duty chisel for knocking out masonry.

Buttering: smearing mortar over the surface of stone etc.

Claw boaster: heavy metal chisel for dressing stone.

GLOSSARY

Club hammer: heavy hammer.

Cob: early form of concrete using chalk, mud and chopped straw.

Cold chisel: sharpened, heavy-duty chisel.

Console: ornamental bracket or support for door hood or canopy.

Cornice: continuous horizontal moulding at the top of an internal wall.

Coving: concave, curved surface between the wall and ceiling.

Distemper: water/glue-based paint, such as whitewash.

Drags: toothed tools for scoring stone.

Droving: parallel, grooved decoration in the surface of stone.

Dummy: heavy mallet.

Entablature: classical architectural term to describe architrave, frieze and cornice area.

Facing stones: outer stones used to create the outer surface of a wall.

Flagstones: slabs of stone used for flooring.

Float: plasterer's tool used to create a smooth finish on plaster.

Freestone: stone which can be worked in any direction, so is easy to cut or carve.

Frieze: upper part of the wall beneath the cornice.

GLOSSARY

Furrowing: deeply grooved lines used to decorate the surface of stone.

Gouge: cutting/shaping tool.

Gypsum: mineral used in the manufacture of cement and plaster of Paris.

Hawk: plasterer's tool used to hold the plaster.

Inglenook fireplace: large, open fireplace, often with built-in seat inside the chimney breast.

Joists: smaller timber beams supporting a floor or ceiling.

Laths: narrow strips of timber used to make a supportive framework.

Lintel: horizontal beam over a door, window or fireplace.

Mortise: rectangular cut used to joint two pieces of timber/moulding at right angles.

Palladian: eighteenth-century style of architecture.

Pargetting: decorative moulded plasterwork.

Paver: formal slab made of stone or concrete.

Pebbledash: external finish for walls where small stones are applied to the damp render.

Pier: pillar designed to bear heavy loads.

Pilaster: shallow, rectangular column on the face of a wall.

GLOSSARY

Pitching tool: type of cold chisel.

Pointing: mortaring the joints between stone, brick etc.

Portico: covered entranceway.

Quirk: tool for raking out pointing.

Quoin: external corner of a wall.

Regency: architectural style during the period 1811–1820 in Britain.

Render: external plaster finish for walls.

Roughcast: pebbled finish for external rendered walls.

Rubble: roughly cut stone.

Scabbling: scooping out sections of the stone's surface for decoration.

Skim: jute sacking.

Spirit level: tool for measuring the levelness of a vertical or horizontal plane.

Spotboard: clean piece of board used for plaster or cement.

Stucco: smooth or modelled plaster coating for external walls.

Waster: chisel used for taking thin shavings of stone.

Wattle-and-daub: early form of infilling for timber-framed buildings, a timber framework daubed with clay and finished with lime plaster.

INDEX

INDEX

INDEX

INDEX